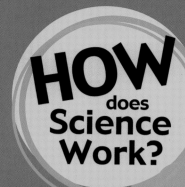

HOW does Science Work?

Exploring Forces and Movement

Carol Ballard

PowerKiDS press.

New York

Published in 2008 by The Rosen Publishing Group, Inc.
29 East 21st Street, New York, NY 10010

First Edition

Commissioning Editor: Vicky Brooker
Editors: Laura Milne, Camilla Lloyd
Senior Design Manager: Rosamund Saunders
Design and artwork: Peta Phipps
Commissioned Photography: Philip Wilkins
Consultant: Dr Peter Burrows
Series Consultant: Sally Hewitt

Library of Congress Cataloging-in-Publication Data

Ballard, Carol.
    Exploring forces and movement / Carol Ballard.
        p. cm. – (How does science work?)
    Includes index.
    ISBN 978-1-4042-4277-7 (library binding)
    1. Force and energy—Juvenile literature. 2. Motion—Juvenile literature. I. Title.
    QC73.4.B36 2008
    531'.6—dc22
                                        2007032024

Manufactured in China

Acknowledgements:

Cover photograph: Airborne Snowboarder, Mike Chew/Corbis

Photo credits: Martin Rogers/Getty Images 4, Lorentz Gullachsen/Getty Images 5, Leo Mason/Getty Images 6, Paul Ridsdale/Alamy 8, Science Photo Library 11, Frank Whitney/Getty Images 12, Larry Williams/Corbis 14, Luca Tettoni/Robert Harding 16, Steve Smith/Getty Images 20, Mike Chew/Corbis 21, Gregg Adams/Getty Images 23, Hisham Ibrahim/Getty Images 24, Roy Mehta/Getty Images 26, Michael K. Nichols/Getty Images 28.

The author and publisher would like to thank the models Alex Babatola, Sabiha Tasnim, and Zarina Collins, and Moorfield School for the loan of equipment.

Contents

Words in **bold** can be found in the glossary on p.30

Forces

Forces are **pushes** and **pulls**. Without forces, nothing would move—everything in our world would be still.

Forces are at work all around us. Every day, forces are involved in everything that we do. It's impossible to imagine a world where forces aren't at work somewhere doing something.

We push and pull things all the time—probably a lot more often than we realize.

There are pushes and pulls going on around us every day.

Pushes make things move away from us. Pushes are needed to close a drawer and cut a slice of bread. Pushes can be huge, such as the push needed to knock down a building. Pushes can also be tiny, such as the push needed to pet a kitten or puppy.

Pulls make things move toward us. Pulls are needed to put on your socks and to lift up a cup. Pulls can be huge, such as the pull needed to haul a broken-down bus. Pulls can also be tiny, such as the pull needed to pick up a pin.

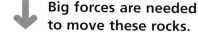
Big forces are needed to move these rocks.

What can forces do?

We cannot see forces, but we can see what they do. A push or a pull can make things start to move. This happens when a ball on the grass is kicked. A push or a pull can also change the shape of things. When you sit on a cushion, you **squash** it into a new shape.

A push or pull can make things stop moving, such as when you catch a ball. A push or a pull can also make moving things speed up or slow down, or move in a different direction. This happens when a tennis player hits a ball that is coming toward him or her.

↑ **The push from the soccer player's foot makes the ball move.**

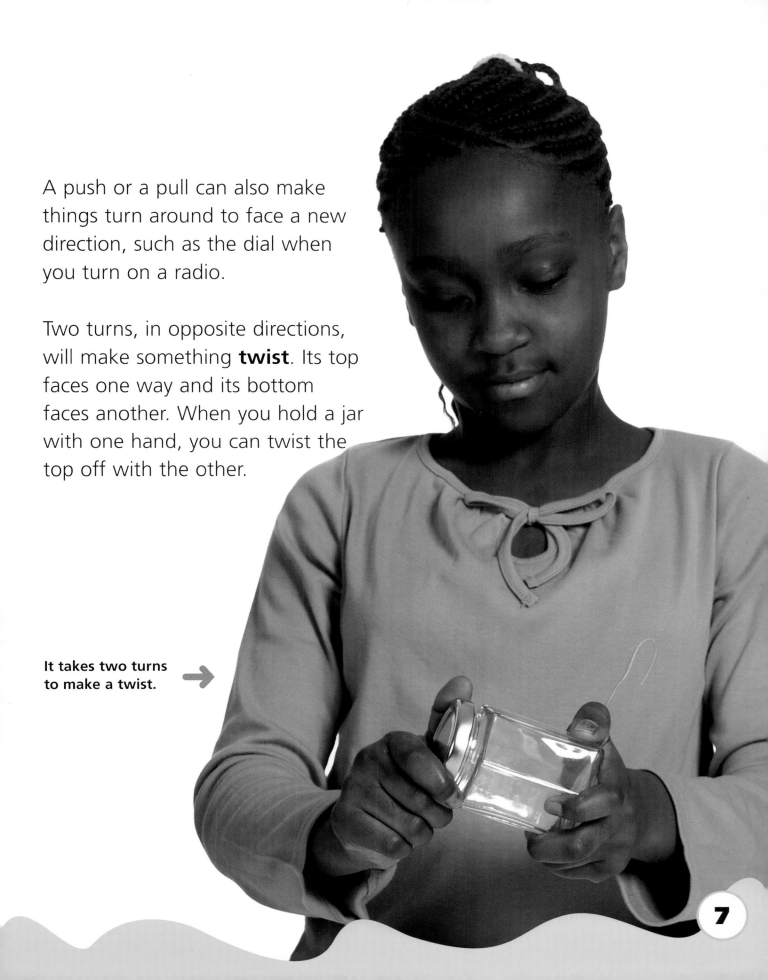

A push or a pull can also make things turn around to face a new direction, such as the dial when you turn on a radio.

Two turns, in opposite directions, will make something **twist**. Its top faces one way and its bottom faces another. When you hold a jar with one hand, you can twist the top off with the other.

It takes two turns to make a twist. →

7

Moving and changing shape

Some forces do several things at once. For example, when a snowball hits a wall or a person, the wall or person's body pushes back on the snowball. This push makes the snowball stop moving, and it also changes the shape of the snowball.

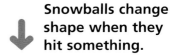

Snowballs change shape when they hit something.

Two things happen if you roll a modeling clay sausage across a surface—the clay sausage will start to move, and it will also get longer and thinner.

Nothing can move unless a force acts on it. Some things need bigger forces than others to make them move. A toy car only needs a small push to make it move. A real car needs a much bigger force to make it move.

Once something has started to move, it cannot change its speed or direction unless another force acts on it.

Forces change the shape of modeling clay. →

9

Gravity

Gravity is a pulling force. It always pulls everything toward the center of the Earth. If there was no gravity, nothing would stay on the Earth's surface—it would all just float away into space.

↑ **When you let go of a ball, gravity pulls it down.**

If you drop something, it lands on the floor. If you throw a ball into the air, it will go up for a little way but then come back down again. If you pour some milk, the liquid flows down. It is easier to run downhill than uphill. All of these things are due to gravity.

⬆ **The pull of the Earth's gravity keeps the Moon in its place.**

Gravity also pulls things that are around the Earth. Gravity holds the Moon in its **orbit** and holds the air in place.

Wow!

When a rocket takes off, it needs a huge force to escape from the pull of the Earth's gravity!

Balance

If something is balanced, its weight is spread evenly around its **balance point**. Tall, thin things have high balance points. Short, fat things have low balance points.

Tall, thin things are less stable than short, fat things. It is easier for them to tip over, because just a little tilt will unbalance them. It takes a bigger tilt to topple a low, fat object.

Low sports cars are less likely to tip over than tall buses.

TRY THIS! Find the balance point

1   You will need some objects, such as a piece of card and a small pad of paper.

2   Try to balance the piece of card on your finger.

3   Repeat this with different objects.

The place where you put your finger to balance the object is its balance point.

Your own balance depends on your balance point. See how you wobble when you stand on one leg!

Spinning around

When things **spin**, a special force keeps them turning in a circle. Usually, things that are moving keep going in a straight line until a force changes their direction. But when something is spinning, it goes around and around in a circle.

All the time it is spinning, a **turning force** keeps it moving in the circle. Heavy things need a bigger turning force than light things.

Turning forces keep these fairground rides spinning around and around.

TRY THIS! Make a spinning force

1 You will need a small ball and a long sock or a pair of tights.

2 Put the ball in the end of the sock.

3 Find a large empty space.

4 Hold the other end of the sock in your hand, and spin the ball around and around.

5 After a few spins, let go of your string and watch to see which direction your ball flies off in.

The spinning force stops when you let go of the ball, so you should find that it flies away from you in a straight line.

! Be careful when you let go of the ball

Measuring forces

Sir Isaac Newton was a British scientist who lived more than three hundred years ago. He spent a lot of time studying forces and how they work. Most of what we know about forces is based on his ideas. Forces are measured in units called **newtons** (N) because of Sir Isaac Newton.

We can use a **Newton meter** to measure the size of a force. A scale on the side of a Newton meter shows the range of forces it can measure.

Forces can be very big, such as the force needed to pull a car.

These elephants need to use a big force to move the heavy tree trunks. →

TRY THIS! Measure the size of a force

1. You will need three unbreakable objects, a light one, a heavy one, and one that is somewhere in-between.

2. Hook the first onto the Newton meter.

3. Hold the top of the Newton meter and let the object dangle freely.

4. With the scale at eye-level, read the size of the pulling force.

5. Repeat this using your other objects, one at a time.

You should find that the heavier the object, the bigger the pulling force.

A Newton meter has a **spring** inside. When the Newton meter is pulled, the spring is stretched. The bigger the force, the more the spring is stretched. You can read the size of the force by looking at the position of the marker on the scale.

Friction

Whenever two surfaces rub against each other, there is a force called **friction**. Even surfaces that are very smooth create friction. As one surface moves over the other, the rough bits catch on each other and cause friction. The rougher the surfaces, the more friction there will be.

Friction works to slow things down and stop them from moving. It also makes the surfaces get hot. You can feel this happening if you rub your hands together hard. Friction between your hands makes your hands feel warm.

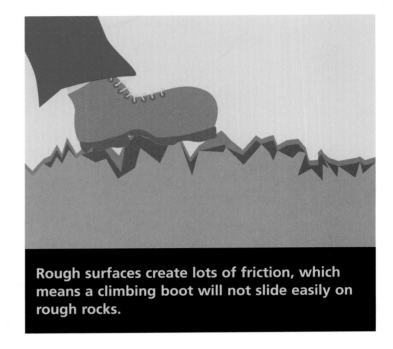

Rough surfaces create lots of friction, which means a climbing boot will not slide easily on rough rocks.

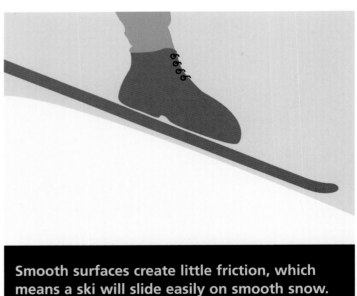

Smooth surfaces create little friction, which means a ski will slide easily on smooth snow.

TRY THIS! Investigate friction

1 You will need to find a toy car and something to make a ramp from, such as a wooden board or a book.

2 Make a slope by resting one end of your ramp on top of something to make it slope downward.

3 Put a toy car at the top of the slope.

4 Let go, without pushing, and measure how far the car travels across the surface at the bottom of the slope.

5 Repeat on different surfaces, such as a carpet, a rug, and a tiled floor.

You should find that the smoother the surface, the farther your car will travel. This is because the smoother the surface, the less friction there will be between the car wheels and the surface.

Is friction useful?

Friction can be very useful when you do not want things to slide over each other. Friction caused by the **treads** in the soles of your shoes stops you from slipping on muddy paths. You need friction between your bicycle brakes and the wheel rims to stop your bicycle. We can increase friction by making the surfaces rougher.

There is friction between something that is moving and the air or water that it is moving through. We can reduce this kind of friction by using very smooth shapes that the air or water can flow around. For example, swimming caps allow swimmers to go faster by reducing the friction between their hair and the water.

The deep treads in this snow tire stop it from slipping on the snow and ice.

Friction is not useful when you do want things to slide over each other. It can slow you down when you ski across snow, or skate on ice.

Friction also makes machines get hot as the parts rub together, and it wears down the machine parts, too. We can reduce friction by making the surfaces smoother.

The smooth surface of this snowboard helps to reduce the friction between it and the snow. →

Many forces

When something is still, all the forces acting on it are balanced. It takes a stronger force in one direction to make it start to move.

Think about a ball that is lying on the ground. Gravity will be pulling the ball down toward the center of the Earth. The ball isn't sinking into the ground, so the ground must be pushing up with a force that exactly matches the pull of gravity.

A ball does not move unless it is pushed or pulled.

If somebody kicks the ball along the ground, four forces are at work on the ball. The force from the kick pushes the ball forward. Gravity keeps the ball on the ground. Friction between the ball and the air will slow it down. Friction between the ground and the ball will also slow it down. The ball will stop when the kicking force and the friction are balanced.

When you ride your bicycle, five forces are acting on you. Gravity pulls you down. The ground pushes up. The force from your legs pushes you forward. Friction between you and the air, and between the tires and the ground, push you back and slow you down.

There are many forces at work when you ride your bicycle.

Falling through the air

Gravity pulls things down, but the air pushes them up! The force of gravity pulls everything down toward the center of the Earth. As something falls, the air pushes against it. This upward push is called **air resistance**.

The size and shape of something affects how much air resistance there is as it falls. If you jump out of an airplane without a parachute, gravity pulls you down to Earth very quickly! Opening your parachute means you take up a bigger space in the sky so there is more for the air to push against—there is more air resistance, so your fall to Earth is slower.

A parachute helps to slow down the fall of these skydivers. →

TRY THIS! Investigate air resistance

1 Cut a piece of paper 6 in. (15 cm) long and 1.5 in. (4 cm) wide.

2 Fold it in half lengthwise.

3 Cut 1.5 in. (4 cm) along your fold and fold back each "wing" to opposite sides.

4 Use a paper clip to hold the long fold together.

5 Drop the spinner you have made and time how long it takes to land.

6 Make more spinners using the same size of paper, but making longer or shorter cuts along the center fold.

7 Measure the lengths of the wings of each. Repeat your test with each of the spinners.

You should find that the longer the wings, the slower the spinners fall. This is because they have more air resistance than spinners with shorter wings.

Be careful using scissors

Floating and sinking

Why do some things **float** in water but others sink? When something is put into water, gravity pulls it down but the water pushes up. The upward push of the water is called **upthrust**.

If something is light for its size, such as a beachball or a pool float, the upthrust of the water can hold it up and it will float. If something is heavy for its size, such as a pebble, the upthrust of the water cannot hold it up. It will sink.

The upthrust of the water makes these inflatable loungers float in the swimming pool. →

TRY THIS! See what floats

1   You will need a bowl of water and some modeling clay.

2   Gently put a lump of clay into the water.

3   Now flatten some more modeling clay into a boat shape.

4   Lay it gently on the water.

The lump of modeling clay should sink because it is heavy for its size. The clay boat should float, because the air inside its shape makes it light for its size.

Pressure

The force of one thing pushing on something else is called **pressure**. When you stand up, your feet push on the ground. They are putting pressure on it. The smaller the area touching the ground, the bigger the pressure.

If you stand on one foot, all your weight is pushing on a piece of ground the size of one shoe. If you stand on both feet, your weight is spread across the ground with the size of two shoes, so the pressure is less.

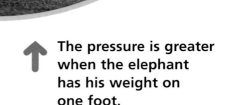
The pressure is greater when the elephant has his weight on one foot.

TRY THIS! Investigate pressure

1 You will need an empty, clear plastic container, such as a bottle or a carton.

2 Make two holes, one above the other, in the container.

3 Stick a piece of tape over them.

4 Fill the container with water.

5 Holding it over the sink or a bowl, remove the tape.

Be careful using scissors

You should find that water shoots out farther from the bottom hole. This is because the water at the top pushes down on the water below. This shows that pressure can affect liquids as well as solids!

Glossary

air resistance the upward push of air on something that is falling

balance point the point around which an object is balanced

float to rest on the surface of a liquid without sinking

forces a push or a pull

friction a force produced when two surfaces rub together

gravity the force that pulls everything to the center of the Earth

Newton meter an instrument used to measure the size of a force

Newtons the units used to measure forces

orbit the path of the Moon around the Earth

pressure the push of one thing against another

pulls forces that move things toward you

pushes forces that move things away from you

sink to fall beneath the surface of a liquid

spin a force that keeps things moving in a circle

spring a metal coil that can be squashed or stretched

squash to put pressure on something and change its shape

treads grooves made in shoes or tires

turning force the force that keeps things spinning in a circle

twist to turn one part of something in the opposite direction from the other

upthrust the upward push of water

Further information

Books to read

Forces and Motion by Chris Oxlade (Hodder Wayland, 2006)

Forces and Motion (Science Answers) by Chris Cooper (Heinemann Library, 2004)

Forces and Movement by Malcolm Dixon and Karen Smith (Smart Apple Media, 1998)

Forces and Movement (Start-up Science) by Claire Llewellyn (Evans Brothers, 2004)

Forces: The Ups and Downs by Ben Craven (Raintree, 2005)

Web sites to visit

Web Sites

Due to the changing nature of Internet links, PowerKids Press has developed an online list of Web sites related to the subject of this book. This site is regularly updated. Please use this link to access this list: www.powerkidslinks.com/hdsw/force

CD Roms to explore

Eyewitness Encyclopedia of Science, Global Software Publishing

I Love Science!, Global Software Publishing

My First Amazing Science Explorer, Global Software Publishing

Index